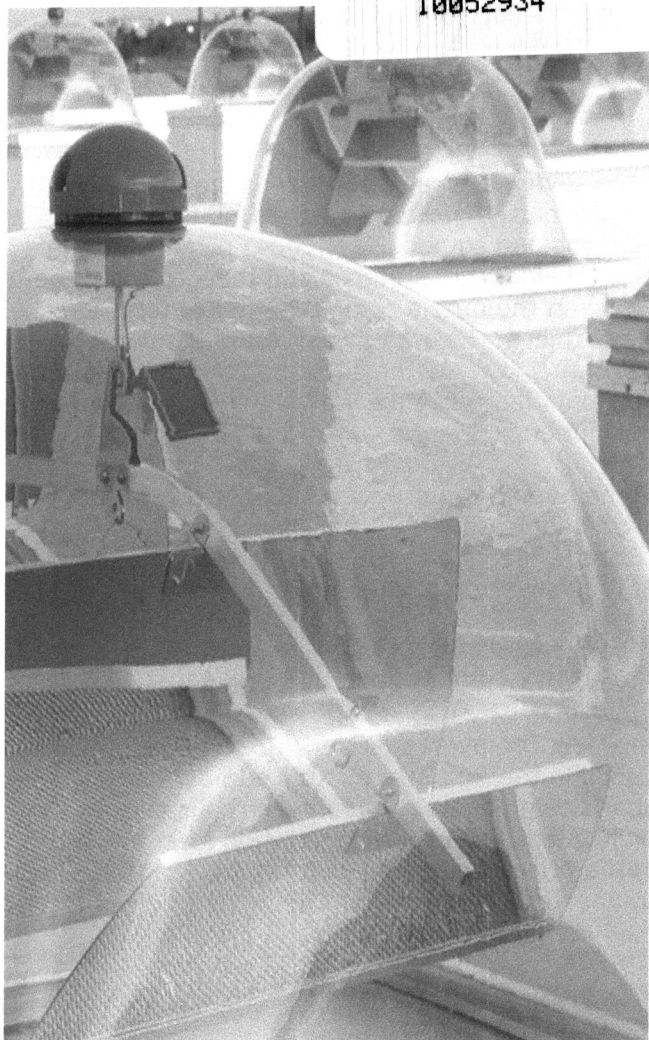

SilveRevolution

I see silver as an algorithm and as a painting.
I study patterns. I'm an artist who creates
patterns and somehow, as those patterns are laid
out, an algorithm surfaces and reflects – and the
pieces in the puzzle come together to form a
Masterpiece. I want to relay some numbers to
begin to show you the silver pattern as a painting.
This painting reflects a War and eventually a Green
Revolution.

Some people can only see silver in numbers.
Most people Can't see it as slavery.
"Experts" see a silver market at $40 billion, or
26,900 silver producing mines.

If you look at the average Comex spot price of
silver from 1975 to 2020, it reads $20.72.

In 2013, the average silver price was $23.79 an
ounce in the first quarter; priced near our 2021
year closing average. It's been reported that the
World Bank's long-term prediction forecasts a drop
in the price of silver down to $13.42 per troy ounce
by the year 2030.

As you take in this relayed information of many years of study, the painting I present will begin to reflect light. In that light, you will see a limited number of steps for solving some major problems:
Our Dependency on Fossil Fuels-
Our Savings Accounts-
Our Trillions in National Debt-
And Our Economy versus China's-

I've read that China sold a third of its silver holdings in 2006 and supposedly they are the largest holder of silver. 95% of China's silver production is the byproduct of other mining. They produce roughly 126 million ounces of silver annually.

The Bayan Obo mine in China is the world's largest rare earth mine. China produced roughly a 57.6% share of the total global earth mine production as of 2020. The rare earths are a group of 17 chemical elements that are used in cell phones, cars, laptops, weapons, TV's, aircraft engines, and beyond.

They play a role in making technologies faster and lighter. There is also an abundance of them in the Earth's crust – and a small amount goes a long way!

In regard to China's exportation of precious metals, I should mention that 70% of China's copper comes from IMPORTED garbage.

Let's examine cell phones as an example of recycling and dissect some numbers. Just in the USA, not including the rest of the world, 75% of our curbside recycling winds up in China for processing – and it employs millions of Chinese workers.

Americans throw away roughly 400,000 cell phones a day!

Let's say at an impressive rate of recovery every 3 days, a million recovered recycled phones are retrieved. A million cell phones salvaged create roughly 772 pounds of silver.

That equates to roughly 121 days per year times 772 pounds of silver

creating roughly 93,412 pounds. When you break that down into kilos,

it's 42,460 and break it down further into silver ounces and you're looking at roughly 1.36 million Silver ounces from USA recycled phones annually. That is a very small fraction of silver produced entailing a lot of Chinese effort!

I've read Russia sold 90% of its silver "holdings" in 2013. Can we Define "holdings"?

Good luck accurately tracing where that 90% went.

I've read that silver is not part of any strategic official government reserves anymore.

I've read a lot of BS from the "experts"!

I'm going to highlight some of that information and challenge it – in order to paint the much broader picture of silver.

I've read that roughly 51.4 billion ounces of silver has been mined on planet Earth throughout history. Some estimates have it at 1,740,000 tonnes or...in other words, all the silver discovered thus far would fit in a cube 55 meters on a side.

A lot of experts voice their frustration daily about not understanding why silver prices are not higher. I'm going to show you that answer in this book. How much silver is out there is a common question by all the experts.

Without going down the list of 26,900 silver producing mines one by one – let's focus on some of the key players that help provide the estimated 25,000 tons mined worldwide – or roughly the 803 million ounces produced annually to supply industries. We can say the fabrication demand is 926 million ounces according to the CPM group.

Mexico is said to be the world's largest silver producer at some 23% of world production and they are reported as producing more than 200 million ounces in 2019. Roughly 6,300 metric tons of silver were produced by Mexico in 2019. So let's just say of the estimated 25,000 tons mined worldwide annually, they are a key player. It will be interesting to see what future Pandemic mining production numbers reflect.

Peru is a key player – lets explore just one mine and learn more about it – This Silver painting is forming an outline. The Potosi mine in Bolivia- is owned and operated by Empresa Minera Manquiri SA, a subsidiary of Coeur d'Alene Mines Corporation of the good ol' USA of course! We've been preparing for a currency war for a long time- ever since the 1965 Coinage Act to be exact! Potosi's plants full capacity was reached in 2008 and is the world's largest pure silver mine - the 15 year lifespan puts depletion and hundreds of miles of mine shafts exploited this decade! I should note that it can take, for new mines, more than 10 yrs before a single ounce of silver can be mined. In true mining fashion, projects across the world are up to 10 years behind the silver race.

How much energy does it take to produce an ounce of silver? Pull out a 2-ton rock, crush it, and retrieve at best $250 in silver. Let's ask a Bolivian slave miner how much energy it takes walking down those tunnels! We often forget about the workers.
Ask our Earth what it takes? Remember Earth? FYI, not too many mines are profitable.

Why in the world would we want to make mining profitable or invest in it? I'd like to think we are evolving. It takes conscious thought and preparation and a lot of time to create a revolution. It takes knowledge outside the box of trading to see the broader picture of human survival on Earth. Evolution.
Examine the ozone layer before and after the pandemic.

Examine Earth's movements after industry pause. Did it affect weather patterns?

How many business flights are booked now? And international flights? Contemplate Fracking sparking earthquakes; The Three Gorges Dam in China slowed down Earths rotation. Contemplate that. Contemplate the Gulf Oil Spill.

When we pollute and degrade our Earth, we destroy our ability to sustain ourselves optimally. Industrialized Farming is an example of supporting pollution –
Mining is an example of supporting pollution. Industrial Trade is another example.
The World Trade Organization's fine print policies pave the way to bypass most environmental regulations. Some exceptions are made with "provable scientific justifications" and acceptance from the WTO. **Policies must change for us to Evolve**.
International Trade in essence provides a favorable avenue for the cheaper polluting producers.
Our standard of living, and our impact on the primary production of the planet, reveals our business practices and policies are destroying life on Earth. Hundreds of thousands of species will go extinct due to our current policies that dictate our world.
At the current rate of extinction, we face a loss of 20 percent of all species on the planet **within** the next three decades!

There was an estimated 90,000 plus hazardous waste sites in the United States in 2010. The most toxic locations tracked by the federal government are called Superfund Sites. In 2017, there was a reported 1,317 Superfund Sites.

You can access information online to find out how close you live to these sites. <u>It's a very</u> sobering and educational search!

We store and accumulate toxic waste because there is no way to render the majority of these substances harmless.
Incineration spreads toxic emissions which contribute to the Greenhouse Effect and Holes in Our Ozone.

The Nuclear Power Industry claims it provides clean and safe energy – yet, "safe, secure storage" of Plutonium requires 200,000 plus years of guarding.

Solar Manufacturing leaves less of a footprint in the broader picture*
Polysilicon for Photovoltaic Panels creates toxic chemicals.
Truth and Transparency*.
It's expensive to process and recycle the Silicon Tetrachloride – but it can be processed "responsibly" – at a cost.

Right now, we are borrowing from the future in a pattern of short-term abundance & production.

We use as much energy as we want to allow new growth with the rationalization that we're not doing anything wrong and we do it with the diffusion of responsibility.

The U.S. Geological Survey has estimated there is "below ground, ready to be mined silver totaling 530,000 million tonnes."

To supplement our current consumption, we are depleting resources that took millions of years to create.
Our unwillingness to examine our limits reflects in Mining Productions.
Demand for industrial use of silver is expected to go up and production is estimated to grow at an increase of 2.98% from 2021-2025 according to the Compound Annual Growth Rate.

Every twenty-four hours, our Worldwide Economy burns an amount of energy our Earth Required 35 plus years to create! One day of consumption equals 35 years!!!

New Resources do not technically exist until they are extracted.

Stop looking at the gold to silver ratio of 74 to 1 and start looking at mining ratios of 8 to 1.

There is a strategic, suppressing, manipulation restraint taking place one mine at a time all around the world.
All of the experts are fixated on Futures, Options, and Derivatives! We are at the mercy of the government and banks for a reason.
A Liquid Market –
Supply and Demand –
Metals are Critical Commodities.
Look at Mine Capacity –
Look at Lack of Margin – Look at Earth's Reactions!

The demand for industries is a key note here.
The slowing of production is important to note.
The timing of new mines and access to production equals Protectionism.
The lack of new investments equals Protectionism.
Protectionism equals War.
Who influences suspending operations on mines or halting production?
People buy into the facades –
The easy reasons – the news clips –

Perhaps it costs more to mine it and it's not worth the battle right now.

A lot of mining productions were slowed/stopped by governments due to the pandemic.
The pandemic is necessary for Earth study.

It's a pause to evaluate what happens when we halt productions.
It's a hit on industries to weaken them strategically before the war
Officially reveals itself.
The mining industry does not need to be profitable to exist, not when governments are involved in their existence!

Look at net exports. The examination of the numbers are designed to be a Labyrinth to distract and employ the "experts" - to throw them a bone and give them something to talk about day in and day out – to create a "Conspiracy".
The Mint sells out strategically, shifting the experts away with distractions like J.P. Morgan's accumulation of 150-800 million ounces
of physical silver over 8 years or...Warren Buffett or the tired old Hunt Brothers. A broken record!

The CPM's analysis or the Silver Institute's studies or the Silver Users Association guide of what goes where.
Accurate accounting of aboveground silver is estimated for a reason-and "studies" are released because unaccountable silver has been religiously siphoned to Reserve Banks and USA Military Weaponry and Storage for more than half a century and likely much longer!

The Vanguard 1 Satellite was launched by the U.S. in 1958. It was the first solar-powered Satellite to be sent to space. In 1954 Bell Labs demonstrated The First Practical Silicon Solar Cell. Back then, they realized the true value of silver. Eleven years later, enter The Coinage Act of 1965.

Go all the way back to 1883 when Charles Fritts created the first solar cell by coating selenium with a thin layer of gold.
Seventeen years later, enter the Gold Standard Act.
Fast forward to 2021 and it takes about two-thirds of an ounce of Silver to produce one average solar panel.

Let me make it very clear here for the folks who have trouble understanding the Art of War and the connection of national wealth and national security. There are central bank, military, and government silver stockpiles. Go calculate our military's arsenal and reserve locations.

Good luck with the numbers!
Between 1945 and 2009, the U.S. built more than 66,000 warheads.
That equates to a small amount of silver needed.
Maybe 33 million ounces there* But you can keep adding*

I'll cut to the chase with one recorded project as an example.

I've read that the Manhattan Project "borrowed" 29,365,193 pounds of silver to make calutrons for the first atom bomb.

The silver was needed to produce coils to make the calutrons' giant solenoids.

It's reported that the War Department eventually withdrew more than 400,000 bullion bars of approximately 1,000 FTO each from the West Point Bullion Depository in West Point, New York.

The first bars were withdrawn on Oct. 30th, 1942 and sent to a U.S. metals refining company in New Jersey.

The plant began casting the bars into cylindrical billets weighing 400 pounds each.

By the time the casting operations stopped in January of 1944, over 75,000 billets weighing nearly 31 million pounds had been cast!

Fast forward to June 1st, 1970 and it is reported that the Manhattan Project Silver was returned. Now take a breath and think about that for a minute.

I'm telling you that there were cylindrical billets weighing 400 pounds each and they cast over 75,000 billets weighing nearly 31 million pounds!

Then, on June 1st, all the silver was returned to West Point – and some was shipped to England. End of story.

The average folk says, Oh – O.K. Bert – Bert – yeah, O.K. Bert!

But I'm saying on June 1st, they drove back the silver equivalent weight of 7,500 midsize automobiles and said thanks for the loan, guys!! If that amount of silver was necessary to build such a weapon – just imagine where silver is stored and how long it's been siphoned;

The Y-12 facility at Oak Ridge operates as a Department of Energy National Security Complex. Their racetracks required extraordinary amounts of copper for magnet windings – They "borrowed" silver as a "substitute". They are located in Oak Ridge, Tennessee.

I've read that 55% of the total worldwide silver is found in just 4 countries on Earth.

Let's talk about the Federal Reserve.
There are twelve regional reserve member banks that assist the FOMC and the FRS with providing "data" and as a matter of national security, the accounting is likely managed in divisions and those

departments are given figures that get entered, and then divided into intricate, untraceable networks, to create the necessary facades of vulnerability in silver reserves.

In other words accounting is not accountable – hence a silver "deficit".

More consumption of silver than supply – but still – we make ends meet.

This is an abstract painting with a soon to be clearer image!

Take a look at a Satellite image of the Grand Canyon.

Can you see gold?

Can you see mining?

Most people can only see images in representational paintings.

They can't wrap their minds around an abstract work.

I've read that there are 6 billion ounces of aboveground gold on planet Earth.

I've read that the 2018 World Silver Survey claims 2.78 billion ounces of silver bullion are stored. It's estimated that a total of 6 billion ounces of silver exists aboveground <u>matching</u> the estimate of aboveground gold existing.

Worldwide, we are likely conservatively producing 800+ million ounces
of silver annually (150-200 million ounces from recycling and the rest
from mining). Future demand could send us towards a billion ounces annually. Examine the silver painting more closely.

Look at different figures outside the GDP – like whether or not the
Amazon rainforest will be able to perform as a carbon sink.
Soy vs. Beef.
Trees vs. Barren Land.
Edible Fish vs. Dead Ocean/ Ocean Currents Ebb & Flow.
The Limits to Growth are key notes to why there is a silver shortage. The growth of industrial civilization needs to decline substantially and more importantly, stay flat, for us to survive on Earth.
Hence…the race to Mars! Hence…a forced Pandemic!
The adaptations to resource without increasing pollution is the battle we face.
Earth will force us to face it!

Seismological data won't measure long term magnitudes. Common sense confirms we are poking the wrong places and it will result in a fault line catastrophe. Oil & Gas extraction by injection of wastewater produced by the extraction back into the ground creates earth disruptions. Period. Short-term growth equals long-term consequences. Sustainability is the solution without increasing pollution to meet resource demand.

It's cheaper to build a solar farm than a nuclear plant.
There are too many people on the planet!

I've read that less than 20% of worldwide primary silver production comes from silver mines.

No one has come along and attempted to take advantage of the depressed price of silver because they don't have the reserves to make such a move to disturb the current system.

The USA keeps that system and pattern going because it has worked and continues to work to allow us to accumulate mass silver reserves. Buy physical silver at a subsidized price and hoard it! Secretly siphon it. Why is that hard to grasp?

The USA has cornered the physical supply strategically in the markets since 1965.

That's just the market side – add the Military, Central Banks and
Depositories, Mining and Refineries, then times it by roughly 56 religious worldwide hoarding years and you will get an estimate that funds **our** Green Revolution. There's not enough silver to make it Global! We'll start with California*

Silver will keep the USA in place as the #1 World Economy.

In the meantime, you use gold as a front and suppress the price of silver for as long as it takes to accumulate the monopoly reserves across the world until you have a nice, tidy, accessible 75% of the world's aboveground supply of silver. Protectionism.

The rest of the world gets recyclables and their own reserves. No one has paid attention to the behind the scenes silver war brewing because gold is strategically center stage as always*

The World Bank continues to plant the gold seed with suggestions of employing gold as an international reference point of Market expectations - Can't get more predictable than a gold currency system for a "fix". Expect the Unexpected!!! Price affects supply and it's an open market operation – for now. Laws of supply and demand without government intervention is the definition of a "free market."

Our government's job is to milk the trading markets and divert inflation out of real goods and into financial assets. We are experts at trading secretly and rigging all markets. We wrote a library on it in **Acts**.

I recently read that the world's stock market caps are close to $76 trillion and counting. There is a silver market at a measly $40 billion at best! Forty billion is a small number in comparison to $76 trillion.

Contemplate how perfect such a small silver slice is, to avoid detection!
Think about what a measly one billion dollars would have done for Bear Stearns. They could have covered their short positions. But our government wanted to privatize Fannie & Freddie.

If you dig deep enough, you will find that Russia was one of the largest holders of Fannie Mae bonds.
When the U.S. government bailed out Fannie Mae, Russia benefited. Back in 2008 a lot was happening around the world. Thirteen years later, that crash is still affecting us all!

Russian State operated Gazprom is the world's largest natural gas company and makes up ten plus percent of Russia's gross domestic product. Gazprom produces roughly 30% of the world's natural gas supply! They also provide roughly 50% of natural gas supply to Europe. Those numbers will increase substantially if Russia has its way with Ukraine!! That is a key note as trade wars unfold over time, and as Russia increases its monopoly status.

China and Russia would like bilateral trades without using U.S. dollars as a transaction and reserve currency- and they are manifesting that scenario through a Belt Road, an approach built and functioning now.

Our world is changing and others are seeking an alternative to the dollar as a global reserve currency. <u>Trials for a National Physical Currency in Digital Form</u> are underway right now in China. This is not Cryptocurrency! It is a digital currency backed by the Yuan and not backed by stablecoins or pegged to a commodity.

This trial run could allow the option to track spending in Real Time and could open the door to attempt to work around Sanctions in the future-by offering countries like Iran, Turkey and North Korea the option to use digital currency as a form of payment. It could create the ability to subvert the power of the U.S. dollar.

The Central Bank of China runs the digital payment network and the digital Yuan also known as e-CNY, e-Yuan, or DCEP, is expected to eventually replace physical cash and is a virtual form of Yuan issued by the People's Bank of China.

Russia and China have a Yuan-Ruble trade currency settlement mechanism in place. This is a game of RISK. Instead of a predictable market crash, it's possible we will eventually see some crazy rises in interest rates and the bond market will get hammered and spark a significant rise in gold. It's a good angle to inflate gold.*

This would likely only happen if our U.S. dollar is under true threat from Russian and Chinese collaboration. It's all about trade position in the global economy. The Pandemic is creating great tensions in worldwide trade.

Brazil and India have voiced a desire for a more predictable currency system. Foreign nations have called for the dollar's role in world trade to be diminished.

It's a long process and until a serious worldwide trade war/collapse surfaces, we will watch metal prices rise and fall. A lot of silver investors will continue to be discouraged. This is a marathon investment. <u>Patience is a virtue</u>.

Onshore bonds comprise about 90% of the US $15.8 trillion China bond market. China's leading investment outside bonds is commodities - that means buying agricultural land that can be used to grow commodities. It also means buying water rights! It also means gold!

Sovereign Wealth funds may be "SAFE" purchasing off the books from the central bank's view-but not from our U.S. Military radar. The International Monetary Fund sees transparency as well.

It's a convenient loophole to allow the foreign nations to put more of their eggs in the golden basket while We hoard more silver.*

We want the world to believe we are suppressing the price of gold.
We want our adversaries to believe they have gold leverage for as long as we can keep printing money and it works in the markets.

Between March 9th, and March 16th, 2020, the Dow fell a combined 7,362 points. Do you remember that?
Grocery store shelves all across America were empty within a week of that drop. "Wear a mask, you'll be fine!" It's called Behavioral Economics. That's Depression 101 – getting the herd numb. "Our unemployment is low." Ask yourselves why another stimulus then? And another one – and another one and another one!?
From March 2020 to September 2020, sixty million Americans lost their jobs. Lasting effects! The last monumental Great Depression
ended in 1940. It lasted 11 years!
We are in the infancy of a Modern Monumental Depression.

In December of 2020, there were roughly 19 million USA cases of COVID reported and roughly 333K deaths. One year later, as this is being typed, there are roughly 49 million USA cases of COVID reported and roughly 775K deaths. Let's run recent numbers to put this crisis into perspective using a rough estimate of 329 million people in the United States.

329 million people divided by 49 million cases equals roughly 1 in 6.7 odds of falling into "case status."

Of those 49 million cases reported, there's a 1 in 63 chance at death, according to the 775K deaths reported thus far with our population at roughly 329 million people. So...if you test positive, there is currently, roughly a 1 in 63 chance at death.

The CDC Director, Dr. Rochelle Walensky, has been reported as quoting: "We have about 90 to 100,000 cases a day right now in the United States, and 99.9% of them are the delta variant." She reportedly also said the CDC is still uncertain how transmissible the new Omicron Variant is and how effective approved vaccines will work against it but that the CDC is "hopeful" that vaccines will help prevent severe disease;

A pattern <u>not</u> consistent with the claimed solution has us repeating the COVID pattern of 2020 with increased numbers, in light of over 55% of the U.S. population now reportedly vaccinated. A Booster Shot Campaign is underway with regular media posts and news broadcasts frequently reciting the importance of Herd Immunity. Meanwhile...the 2021 COVID pattern shows continued increases in COVID cases and deaths with the challenges variants produce. Johnson & Johnson stock up .42 cents at $163.72 on 10/22/2021 as the FDA nears approval of mixing shots for boosters.

And then you have the illusion of business as usual.........

In March of 2020, the price of Bitcoin was $6,206.00.
In October of 2021 – (10/20/21), it stood out at $64,713.60 with SEC approval of a Bitcoin Futures ETF. In layman's terms, Futures bet on <u>expectations</u> in price moves, for assets such as commodities, stock indexes, and bonds. The leverages amplify the <u>potential</u> gains or losses. I went online in search of a definition for Bitcoin. This is what I found: "It's a data chain of records stored in forms of blocks which are controlled by no single authority."

Bitcoin seems based on perceived value – just as our paper dollar is. Bitcoin has recently been used as payment for certain accepted real estate transactions. A house of cards! Cryptocurrency, <u>as is</u>, has too much room to fluctuate,
and that is likely due to maintaining the perception of a consistent dollar against uncertainty risks a digital currency strategically displays.
Bitcoin could be used as a Black Swan – and that's because the clients being sold this pitch don't have the same opportunities!
You need clients around the world who have risk in their balance sheets and somehow Bitcoin can neutralize that risk if it reaches <u>X amount</u> and ultimately offset their problems. You need to compel greedy clients.
Timing becomes very tricky –
There are a lot of folks needing to offset their problems today! Enter Bitcoin!

Now with a Futures ETF in the equation, when I see ETF, I see MBS' on a Monster level – <u>Our New PRETEXT</u>.
The excuse for yet another collapse – But this time Greed is so inflated there can't just be one Fall Guy, as in Future ETFs.
It's sharks eating sharks on a much more sinister level.

There are countries that will be taken over in the process.

How many banks died in the 2008 collapse? This time – it's countries!!
Constraint then Consolidation. A Pattern.
It can be applied everywhere!!
Let's just use Mining as one victim example.
Mining goes offline. Government Induced.
Certain mines begin to produce more all of a sudden.
Their stocks go up –
Weaker mines get eaten up.
People fall off in the process –
New players surface.
The stakes just keep getting higher.
There is infinite money to play with.

The patterns show that GDP will decline worldwide.
That obviously means trade will take a negative hit across the board.

It's happening right now worldwide.
The USA is on course to spend and print as much inflated, worthless cash as we can, before someone pushes the magic restart button.

It's a button as dangerous as a nuclear bomb order and more deadly. That's why no one has made a move yet!
Right now, it's just Flash Crashes triggering Stop Losses.
The Robot Systems work! The failures of hedge funds and their lender banks create a domino fall pattern. I saw this pattern forming prior to the Pandemic. It started to surface in mid-September of 2019 with sudden money printing. We were facing a financial crisis in September of 2019! Dejavu on a Grander Scale.

Desperate times call for desperate measures.
Enter the 2020 Pandemic! Think about the deaths and SS savings.
The Feds September 2019 balance sheet of $3.8 trillion increased all the way up to $7.2 trillion by May of 2020.

That shows an increase in printing to keep afloat to avoid hedge funds and banks from falling.

We've been playing this song and dance for thirteen long years in an attempt to dig out from the 2008 collapse.
Another collapse is inevitable. Do we really want to play Russian Roulette and create a deeper Global Depression?

Do we have a choice at this point?!?

We are in a War that never mentions silver!
They're quick to underline oil, uranium, copper and gold.
They're quick to publish a lot of books on "precious gold".
Look at gold history and apply it to silver in our modern age*
Franklin D. Roosevelt's Executive order 6111 banned the export of gold from the U.S. except with the approval of the Secretary of the Treasury.

Think about that order and adapt it to fit a silver equation.
Reading between the lines of <u>Orders</u> & <u>Acts</u>, you find more pieces in the patterns.
In 1985, The Liberty Coin Act was approved and it authorized the Treasury to Mint and issue silver coins. It specified that silver would come from national stockpiles. Silver could easily have been on the books as received from The Defense National Stockpile Center.
It's yet another angle of showing the "liquidation" of our nation's stockpiles. BS Liquidation!

Past Presidential administrations who vocally sought to liquidate part of the nation's supply of silver helped ignite future traders to unload future contracts and ultimately kept the value of silver depressed.
We can use the current Pandemic as an excuse to continue the pattern of depressing silver.
As always, allowing the U.S. to hoard and stockpile silver at a discounted rate. Hedge long - take a short position and rake in physical silver.
We need every ounce we can get and it shows at the Mint! One of the most popular silver investment options are the American Silver Eagle coins.

Let's look at the very small fraction of silver coins that have been sold between 1986 and 2015.
Thirty-year Mintage figures show roughly 400 million ounces were sold between that time.
It's a very small slice of the accessible silver market sold to the public if you think about it!
Branches of federal agencies whose purpose is to store and sell raw materials play the role in supply and demand. No rush in unloading stockpiles!
Hence a history of fluctuating annual Mintage numbers.
Let's look at a few Mintage # examples throughout history.

In 1986, the <u>combined </u>Bullion & Proof American Silver Eagles had a total mintage of 6,839,783 ounces.
In 1996, the <u>combined</u> Bullion and Proof American Silver Eagles had a total mintage of 4,103,386 ounces.
Down 2,736,397 from the 1986 totals.

Eleven years later, in 2007 the combined Bullion, Proof, and Burnished American Silver Eagles had a total mintage of 10,471,128.
A small increase of 3,631,345 compared to 1986 totals. Keep in mind, with a new Burnished product in play.*

Three years later, in 2010, the <u>combined </u>Bullion and Proof American
Silver Eagles shot up to a total mintage of 35,624,500!!!
The BP Oil Spill happened that same year and our economy was still suffering from the 2008 market collapse.
We also had extremely high unemployment with 44 million Americans on food assistance.
But hey, in three short years from the 2007 combined total of 10,471,128 Mintage figures- We find ourselves at a healthy 35,624,500 total mintage ounces!

We are talking about Rocky Roads economically and yet an increase in Mintage of 25,153,372 from 3 years prior!

I would argue that <u>this particular example</u>, in an increase this substantial, was not due to public demand!

Silver had a lot of swings in 2009 and prices ranged between $10.51 an ounce up to $19.00 per ounce before averaging out at roughly $14.66 an ounce.

Between 2008 and 2015, silver lost 60% of its value twice!

Yet...production at the Mint increased substantially.

Silver dipped down to $14.27 an ounce in September of 2015.

So what's my point?!!

Examine the patterns and watch the justifications.

There are a lot of lines to read between!

Collectors bought nearly 370,000 of the 2011-W American Silver Eagle Proofs in the first six days of sales, <u>before sales slowed</u> and leveled off.

Those Proofs have an estimated total Mintage of 947,355.

The total 2011 Bullion Mintage went up to 39,868,500 ounces!

Fast forward to mid-January 2013 and the U.S. Mint <u>temporarily sold out of 2013 American Silver Eagle bullion coins.</u>
Sales were suspended until "they could build up an inventory."
The total Mintage <u>just for</u> 2013, Bullion Strike Eagles, was 42,675,000 ounces.
The year 2013 set a record for yearly sales.
The Mint sold 258,860 2013-W, Proof coins in the first five days of ordering and in the same breath, the Mint announced "customer demand will determine the number of coins minted."

No one is going to exhaust the Central Banks of gold or silver reserves! Rest assured.
With a total Bullion Strike Mintage of 39,868,500 in 2011, and a decrease in 2012 Bullion Strike Mintage of an estimated 33,742,500 – ask yourselves why, in all of the months to declare, WHY, in mid-January of 2013, would the U.S. Mint inform its authorized purchasers that it had temporarily sold out of 2013 bullion coins?
Ask yourselves how many collectors are there?!
Why mid-January?

More importantly, how many authorized purchasers are there?

Over Twelve? Who do they unknowingly purchase for? There are over 1,500 dealers on the exchange.
Remember this is a very small fraction of the silver pie!
The bullion market – that is!

Great extremes are being taken to manipulate pricing and to fix the books.

I'm not convinced public demand is fueling the Mintage "increases"!
Meanwhile, dealers across the country have a silver shortage.
I'm making a contradictory statement here.
Dealers could argue there is intense demand! I can tell you they won't sell their inventory with a significant price drop!
If a dealer has sufficient access to supply to meet demand, why do they sell out when the shit hits the fan in a week? Why can't I get the product I want Now? "Branches of federal agencies whose purpose is to store and sell raw materials play the role in supply and demand."
In other words – We need every ounce we can get!

On February 20th, 2014, due to an "inaccurate sales report", the Mint announced an offering of almost

4,200 2013-W Proofs in their inventory to 780 buyers whose orders were canceled.

Look at Demand from a Collector's perspective using the 2013-W Proof example of 258,860 coins sold in the first five days of ordering and a rough estimate total of 934,331 Mintages for those Proofs.

In the same breath, the Mint announced "Customer demand will determine the number of coins minted."

Let's examine how dealers get provided Mint coins to sell to the public.

Bullion coins are bought in bulk by Mint-authorized purchasers.

Participating banks get in on the action as purchasers – along with brokerage companies, and precious metal firms, etc.

Authorized purchasers have to have a tangible net worth of $5 million to be a silver purchaser.

The purchasers sell them to secondary retailers, who in turn, sell to the general public. That's a lot of fine detail information to take in.

I examine patterns.

I question patterns.

I am questioning supply and demand!

I am questioning who, at the end of the day, receives the bulk of the silver! There is a very small market of tangible silver investors. The proof is in the Proof Collectors.

Somehow, in extremely difficult economic times, I examined a change in the pattern and a sudden increase of 25,153,372 ounces surfaced in 2010 - over a 3-year period increase.

That increase broke the Pattern and created a new Pattern.

2008, 2009, and 2010 were tough years economically, Period.

Hardly a time to be rushing in to buy with a minuscule average increase of roughly $1.50 an ounce leading into 2010's pattern changing increase in Mintage of 35,624,500 ounces compared to 10,471,128 ounces in 2007.

Let's not forget the patterns.
Let's not forget 2007 leading into 2008,
...the "good times" leading up to an attack on every stock, bond, and commodity in the world.
Inflation bubbles in commodities – a pattern.
Wires under the sea – a piece.

Space exploration – a pattern.
"Affordable Housing" funding – a piece.

High margin applications – a pattern.
Hedges & "Accountability" – a piece.

Look at $50 trillion in destroyed wealth in the months following the Panic of 2008.
History repeats itself. In 2008, while the market was collapsing, the price of silver was cut in half. Gold lost 30% of its value.
In March of 2020, silver was at $14.00 an ounce and no one could buy it! Not in Bullion! You could pay double spot and get certain product. Now, it's not uncommon to pay $7.00 to $12.00 over spot for bullion.

Silver drops in price when there's a shortage of physical metal – Go Figure! It's a liquidity squeeze. There's strategic demand and it's not fueled by collectors. When you cannot get specific physical product from dealers, wholesalers, refineries, or the Mint, that's strategy at work.
Just part of the War, Forty million ounces from the Mint, Pocket change!!!
What is available now for tangible purchase? Price affects supply! The numbers are whacky!
What do you have to pay to get it?
Can you buy what you want?
None of it adds up correctly.
We've seen gold double in value since 2008.

But what about demand?
Note: The World Gold Council said global demand for gold during the first half of 2021 was the lowest since 2008!
Think about that in relation to the timing I mentioned regarding the sudden increase in Silver Mintage in 2010.
A pattern changer!
Note: Less than six years ago, gold was at $1,050.00 per ounce.
In August of 2020, gold was $2,000 per ounce.
We mine for gold to get copper and silver.
Remember in 2010 and 2011 when people were stealing copper all across our country?
You start to see that when there is a shortage in metals needed, different extremes happen on Real Street! But on Wall Street, Gold will cause disorderly markets.
That's gold's role now – Trade balancing or Disorderly Markets. Safe Haven Illusion.

In April of 2016, China announced that it was establishing a Yuan backed by gold.
The International Monetary Fund (IMF), accepts the Yuan into the special drawing rights.
This means it's accepted as a global reserve currency.

The IMF has cleverly increased the role of the Chinese Yuan – with the help of U.S. voting power in the IMF.

That's an important note in this Chess Match - providing the illusion that gold has relevance and setting the trap to continue to deter the world's view of silver's importance and keep gold in primary demand. It's strategic.

Let's broaden the picture and explore some of our alliances and how we helped "nurture" them with gold, Namely Germany and France, (An odd couple).
England and the U.S. collaborated in 1944 to create a new world monetary order.
It was called The Bretton Woods System.
It's funny how alliances come into play!
Keep in mind Europe was at battle with the German empire in 1916. We joined in shortly after.

Time heals and trade helps!
The Bretton Woods System worked from 1944 to 1973.

The final chapters of this era began right around the time The Coinage Act of 1965 came into play. Keep in mind the escalating costs of Vietnam around the same time.
A good amount of sacrificed U.S. gold ultimately ended up moving from the United States to its export partners during that era,
Namely to Germany and France – now, also known as the ECB.
France was fixated on gold.

It was likely a strategic alliance move with the U.S. and used as a catalyst for "balanced" trading.
Gold became more of a necessary move by the U.S. to appease its alliances and share the gold to deter the world's view of silver's importance and keep gold in primary demand*

(Sound familiar?)

The world bought in hook, line, and sinker – just as they always do. It's the Gold Doctrine.
Gold's role as an automatic "rebalancing" mechanism in trade safely keeps the interpretation of gold's value alive. Right where we want them.

Now we know the Chinese road to Russia is built - and expanding.

China and Russia believe they will corner the gold market and be Powerhouses in creating an alternative reserve currency backed by commodities for international trading.
The USA cornered the silver market a long time ago and we have plenty of gold!

Go investigate the Commodity Futures Modernization Act;
Check out details regarding the Exchange Stabilization Fund;
Watch the Fed print dollars and create higher inflation in China. We will continue to print more because the U.S. debt gets devalued so foreign creditors get paid back in cheaper dollars.
It's O.K. if it eventually results in us receiving much higher cost imports from China – because the inflation also eventually means higher prices for metals.
It also means the USA devaluation will cause higher unemployment in China. A Win-Win!!

We intensify the stealing of China's growth.
Debt leads to Deflation and Inflation breaks deflation. Either way you
look at it – China pays!
China's GDP compared to the USA in numbers is $6.7 trillion less than the American GDP.

I've read that China's economy <u>could</u> surpass the U.S. around 2028.
Do you truly believe that the USA is going to allow that to happen!?
Hence...A Wuhan Virus Pandemic and a Global Depression!
History Repeats itself – Sorry!
How do you think the Industrial Revolution started? War sparked it!
How do you think our Green Revolution will occur?
How do you think the USA will maintain its status as the #1 Economy
in the World?
We will grow our way out and make our national debt sustainable.

A revealing of our Silver hand now is necessary to spark Evolution.
Pandemic/Recession/Depression/Deflation/
Inflation/War/Revolution/ Evolution.
It's a pattern. You could view it a number of ways.
The Fed is the CIA.
The CIA is the IMF – aka: USA controlled Global Central Bank*
We are Machiavelli- 2PAC Style*
We broadcast Breaking News!
"Inflation has arrived with a vengeance."
Chinese exports are causing this or that –
Port shutdowns – New Variants – Cyber Attacks –

You get the idea – Fill in the blank.
It's a story – Deflation – Inflation - Rollercoaster ride.
Robots Buy – Robots sell -
Markets Down –
Job reports Bad –
Oh wait…, 3 days later…it's good*
Markets up –
Robots Alert. Dollar down – Dollar up –
Trade conflicts, petty sanctions, hurt feelings,
HAARP brings erratic weather.
Yada, Yada, Yada.
It's part of the Silver Painting and part of the War,
Until the story changes!

Nanotechnology could evolve to provide for all things silver.
It's possible.
Newly discovered mines could open up billions of ounces of silver. Eventually.
New technology can change the game. Inevitable.
In 2015, Planetary Resources launched a spacecraft to test technologies designed for asteroid prospecting.

In November of 2015, Congress passed the SPACE ACT of 2015, which grants citizens the right to sell and mine material from outer space.

All we have is Now and right now even if billions of new ounces are discovered, we still have a silver shortage.

We still have an Earth that reacts to getting poked! Even if we find new cost effective ways to locate and produce silver, we face the realities of what it does to our Earth and our people.

Based on my studies of mining production, I have yet to find an earth-friendly, people-friendly model–methodology.

So…we may have to "evolve" and go meteoroid mining in outer space.

Until then…we face the realities of needing more silver.

I read a recent report from a "commodity analyst" that declared industrial demand of silver makes up 50 percent of the silver market.

Let's play with some hypothetical SHITuations – There may be contracts in place with American suppliers and American alliances that say – We will deliver X amount of silver to somewhere in the world – and somewhere in that contract, if we or our alliances do not deliver, then there is a cash penalty.

Elements are not principle protected.

The businesses that need their silver will be shorted all across the world!

Derivative contracts with clauses that state you must take cash if there are international trade disputes will receive inflated money instead of silver.

Meanwhile, a lot of people with cash are on the sidelines wanting silver and only able to retreat to gold!

Gold could be used to mitigate a crisis – and create one – it's a clever way to herd investors into ETFs to "protect" their portfolios with gold.

People will Panic Buy – There won't be any exit markers!

It will truly be leased gold owned by central banks. Disguised "interest bearing gold."

Let's just use 2016's estimates of roughly 5,000 ETFs with a value of close to $3 trillion dollars.

There will be investors on the sidelines wanting silver and only able to buy the illusion of gold backing. It's a strategic detour you can't get around. The design has been mastered. Recycled hedges – smart beta – risk parity – value at risk – paper trading –

The procedures' fine print will force captive investors to be exposed to market movements and no one will access physical bullion but the Central Banks! But this is too predictable!
The bullion the investors think they are backed by will be sold for U.S. dollars <u>during the exposed market movements</u>. The herds' portfolios will crumble and the bullion will go to the central banks. Market wise, gold is the focus. Industry wise, there are going to be a lot of disputes.
All those clauses are going to surface simultaneously. Central banks will start withdrawing liquidity and demanding all tangible silver!
It will be happening while the focus is on gold.
There won't be a price-silver-flow mechanism.

There won't be an Exit strategy!
It will be a Historic move, and this time around, there will be zero
mercy. Open interest – silver futures – Rules out the window -
We will be taking delivery.
A small note: Back in 2016, the SLV held over 350 million ounces of silver bullion. The Delivery Demand I'm talking about will be a demand never witnessed before on a global level.

Industrial users of silver will feel it instantly.

Shock and Awe!
Future contracts burnt.
Worthless cash paid.

People will be herded towards gold – forced in – no escape out!
Industries that need silver will adjust and be forced to pay the premium.
It's a good excuse for a market correction.
Consumers will pay more.
Massive trade disruptions.
Extreme tariffs and severe trade sanctions, including embargoes.
Remember the 1973 oil embargo? Apply it to silver.

If we are approaching a more serious World War, I could see the USA,
in partnership with our alliances, doing exactly what China did in 2010,
when China announced a 72% reduction in rare earth exports to Japan.
But that was between China and Japan then.

This time around, the USA does it in alliance with the major central banks – A $olid $even.
The FED, ECB, BOE, BOJ, SNB, BOC, and RBA.
And yes, I said Japan!!
The Art of War – in modern language –

You can bet the enemy will come and you damn well better be invincible when they do!
Don't forget our USA, Ace up our sleeve, Global Central Bank: The IMF.

We bailed Japan out of their nuclear disaster in 2011 and that bailout comes at a very hefty price for Japan and the World, especially when Japan dilutes their stored toxic barrels and pumps their waste back into our ocean starting in 2023.
Fuckashima!
Or...as the Koreans would say: Shinbyong.

The Tokyo stock market dropped 20% in two days back then.
Go do some research on that while you savor some tainted fish – and remember, if we have to align with the enemy, Japan will be in the equation and lots of poor people will die all across the world.
The price of metals will go up.
Foreign countries will tap into their silver reserves.
There will be a global scramble for commodities.
In April, 2020, the U.S. Commerce Department reported a $49.4 billion U.S. trade deficit. That report doesn't hurt us much because our exports make up maybe 4% of our GDP.
That means our exports fell more than 20% and that means imports fell too!

We wanted to feel out what 10% in lower imports feels like and the at-home repercussions.

World trade is going to be in big trouble as this war unfolds.
China invests in U.S. Treasury bonds to keep its export prices lower;
China only owes about $1.1 trillion in U.S. debt.
Keep in mind that roughly 39% of our total national debt is owned by foreigners – not just US taxpayers! (An often unmentioned key note).
Also keep in mind that deflation increases the real value of debt – and that decreases Global Growth.
Nothing compares to silver appreciation in an Inflationary environment – but Deflation, here we are!!

Do you notice what's happening to silver in this deflationary environment?
It's stable.
Go figure!

That's a new pattern forming.
So what will happen in an inflationary environment?
It will increase again*
Win-Win!

Deflation is happening right now with Decreased Global Growth. In 2017, China's trade surplus with the U.S. was $275 billion. The free-trade game is officially changing. More restrictions are coming into play. An important note regarding gold and a potential global monetary reset. <u>A very Big IF!</u> If China and Russia spark it, Iran and Turkey would be along for the ride - We may allow them to press the button and briefly change trade routines (highly doubtful) – But we will counter punch the design and whatever unfolds could morph into a trade currency pegged to gold. That would have to happen after the global markets go haywire and gold is heisted.

This modern digital age transition in essence could become digital ledgers.
Trade currency with the illusion of physical "backing". Moving around physical gold to settle balances likely would happen <u>without</u> physical transfer. It would be a distributed ledger monetary system that uses invoices/ Digital tokens of "gold" –
The exchanges being invoices and tokens - for world trade initially.

This new system could literally turn the cards and make gold <u>contained</u> with little volatility – but with an illusion of free trade – maintaining the peg. (Kind of like what they've been doing to silver all these years – But apply it to gold – with a new set price that doesn't fluctuate much).

It would be an "open-market" operation controlled by the USA, IMF & ECB as the main driving forces. A refined, rigged international monetary system, with a more precise level of victimization worldwide. The new system would be designed to control all exporters and ultimately the world!

Almost no gold will be needed to participate and smaller fish will be forced to invest in the assigned economies linked to the control of the digital currency – namely the $olid $even Central Banks and the IMF. China and Russia will be forced to comply.

Our alliances and U.S. will provide the "gold" tokens for settlements through digital currency transactions. A gold standard at a <u>fixed</u> exchange rate.

Just to provide a broader view on why the world will follow U.S., in the spring of 2018, the managing director of the IMF said global debt

stood at $164 trillion!
Excluding financial institutions, global, private and public debt in 2015 totaled $152 trillion.

So...in three short years, global debt shot up 12 trillion dollars. That's a global pattern that has deep history. Since 2002, the global debt has more than doubled!

The entire world economy is very close to revealing – it's all based on lies. Losing most everything beats the message in. Fiscal dysfunction sums it up – worldwide. BS investment instruments based on futures.

Massive property bubbles just waiting to shatter. Reckless quantitative easing and negative interest rate policies.

Does it mean it will result in the economic collapse of the entire world?? I'm a Romantic; I'd like to hope not!

The long list of U.S. scheduled funding for our Military, Infrastructure, Affordable Housing, and on and on, will eventually push us towards hyperinflation, and in that, we will create extreme winners and losers and we will begin to rearrange economic relations and preserve our government wealth in the process. Note: Government Wealth. But first, Deflation.

This should place silver at its $14.00 range per ounce given decreased global growth and a Pandemic – But it's held strong.

Even with a Flash Crash on August 9th, 2021, with gold dropping $60.00 in minutes and triggering stop losses, silver held strong and recovered. That shows a new pattern is forming regarding Silver.

With new patterns forming – new consistency surfaces and sticks.

Just as we rode a $5.00 to $9.00 bottom line ounce ride pattern mostly through 1986 all the way to 2005, the new pattern surfacing will show a five-year versus twenty-year stretch.

It's very likely we are currently at our new pattern lows at $20.00 to $25.00 an ounce, which means you are technically buying in at $25.00 low in 2021 – as if you were buying in at $5.50 in 1986.

A 20-year ride becomes 5 years, which means, in 2026, the pattern

increases again. As the realities of our mining issues reveal themselves, (Notate Global Mining Output), I believe rushing to sell at
Peak this time around will permanently leave most people out from being let back in at these historic lows – in regards to physical bullion silver.

The monumental drops are designed to weed people out – and the new patterns show that even in a drop in Feb-Mar 2020 to $12.00 - $14.00 an ounce, no one could catch the buy-in for physical bullion.

That means, even though it dropped to those levels, you could not achieve buying it anywhere near spot!

I remember shopping for deals at $14.00 an ounce and finding limited product at double the price – back in March, 2020.
My suggestion is to sit relaxed with your physical silver in the plummet. Educate yourself on Global Mining Output and investigate active full production mining numbers.
It gives you a clearer barometer on the realities of availability.

Simple online searches of what products are available can give you insight on demand.

Right now, I cannot find my preferences for purchasing. People are limited to what is in stock. That's not a favorable position as an investor. It can lead you towards a counterfeit option. Silver coating – lead filling. This new low we are currently in won't ride out 20 years like the old pattern.

So referring to old models for reference points is no longer in the equation. I place our new attainable low at $20.00 an ounce from this day forward in 2022, meaning you likely won't be able to obtain an ounce of bullion silver for less than $20.00 ever again!

It's a pattern low we will look back on, wishing we would have bought in at a perceived "middle market", but in all reality, it is our final lowest tangible "buy-in" since 1986! I'm talking physical bullion here! I believe the new 5-year pattern I'm examining versus the 20-year pattern I referenced will shrink substantially over the upcoming years.

Strategic deflation followed by hyperinflation in the USA, will force foreign countries to show their hands regarding their silver reserves between now and 2028, or before China takes a #1 Economy role.

More importantly, foreigners will be forced to use those reserves and/or face a collapse of world trade.
Protectionism is War. We are at War.

The USA will emerge more powerful than before because we have been buying hard assets with our worthless cash and we've been hoarding silver.
It might keep on dragging out to some level incomprehensible!

Too late, we're here!!
The system has infinite money.
High Frequency Trading generates over half the general stock market's daily volume.
Phantom Shares, Dark Pools, Micro Pennies in Nanoseconds, Flash Crashes, BTOs, CDOs, and on and on.

Don't Buy Futures and Don't swim with Sharks!
Buy Discounted Silver Bullion <u>while you still can</u>.
Store it yourself. A bird in the hand is worth more than two in the bush.
Double your investment eventually.
Be grateful for double your money on an investment.
Even if you hold!

There is a system in place that will not allow silver to skyrocket
because silver will fuel the USA LED Green Revolution.

I know it's a hard pill to swallow after the many years of being told silver will explode.
Even at buying in at a perceived middle, at $25.00 an ounce for silver, and $7.00 to $12.00 above spot price, that makes silver purchasable between $32.00 and $37.00 a tangible ounce.
In rare opportunities, you can purchase in quantity at $30.00 an ounce.
That's buying in the middle at this stage in the War.
However, I see it
as a Historic Low!

In 2013, the average silver price was around $23.79 in the first quarter. We are priced at an average from nine years ago and a lot has changed!

I recently did an online search and read that, overall, silver demand was down 10% in 2020 due to COVID and lower industrial demand.
The spot price fell as low as $12.00 and ounce.
Yet...there was a global deficit estimated at 126 million silver ounces in 2020 – and the spot price

has at times increased upwards towards $26.00 an ounce.
Show us the overall demand numbers for 2021.*
Let's use that 10% drop in 2020 as our barometer to call Bullshit!!

The U.S. will not show its Silver Ace card until a Green Revolution reveals itself.
Or…until more people read and share this book.
Ask yourselves, will the U.S. allow our foreign competition to "Get rich" off gold!?

How about a very small fraction of American investors?
The answer is simple – briefly for the trap.
The one percent that time it perfectly will benefit – prior to the trap.
The 99% remaining will get screwed.
The Gold Doctrine.

Gold's purpose, at best, is to mitigate crisis or create crises!
Nearly all the gold that has ever left the ground is still in existence.
Gold could serve more of a purpose as innovation in science advances* This includes, but is not limited to, the medical and solar sectors.
Touch-sensitive synthetic skin made utilizing Silicon and gold exists today – and the sensors in this new

technology can pick up on temperature and moisture.
Advanced future developments of this particular technology could eventually create the ability for people with prosthetic limbs to feel sensations via artificial limbs.

Gold could become essential in our modern transition and further expand its function. From a grand manufacturing standpoint, gold would obviously need to be priced lower for a monumental industrial role.
In regards to silver, new manufacturing of solar panels is in development and cells are being created closer together in an attempt to cut down the amount of silver needed. As this new design becomes fine-tuned, we could see a panel surface that uses much less silver than the current average.

There are more than a billion humans in developing countries who live without electricity. An additional portion of the world's population endures limited power and frequent blackouts.
With roughly 7.9 billion people on the planet, moderate estimates put the odds of <u>regular access</u> to electricity at roughly 3 out of 4 people on the planet receiving the luxury privilege.

If billions of people are to rise out of poverty, emissions in the developing world will rise with them. Renewable energy <u>needs</u> to be their leading source for power or fossil fuels will continue to have a negative impact on our climate.

Wealthy nations such as the U.S. realize funding the poorer countries' necessities is risky business and it reflects in lack of investments for foreign-cleaner technology projects from private-sector investors.

Astoundingly, renewable energy accounts for nearly a quarter of the world's power!

The delay to create a much higher percentage comes down to wealthy nations' dictations and profit allocations. The Green Climate Fund and The Global Environment Facility utilize the U.N. to channel funds to the developing world – using funds to leverage private capital and contributing a small amount of future pledges could sum up this oppressive pattern.

Private-Sector Financing vs. Wealthy Nation Climate Funding.
Grants vs. Loans.
Less Debt on Developing Nations vs. Wealthy Nations' Greed!

Solar Farm Sourcing and Leasing vs. A Chance at Free Electricity when your Panels are paid off!

The ulterior motives explain the continued delays.

In regard to Battery Plants and Electric Cars, it has been reported Stellantis and GM plan to spend <u>roughly</u> $35 billion <u>each</u> through 2025 on battery plants and electric vehicles.
Other auto makers don't seem to be as ambitious with their spending.
Toyota CEO Akio Toyoda has been quoted stating: "We can't forget that carbon neutrality is also a jobs problem."

It's been reported that Apple Inc. is interested in diversifying toward the vehicle manufacturing business and has aspirations to produce its own car.
Foxconn Technology Group is the world's largest contract electronics manufacturer and they are working to build automobiles for other companies. Their product platform could be used by other brands interested in electric-vehicle models using Foxconn prototypes. As this unfolds more over time, car makers will need to look toward batteries to stay competitive.
That means more demand for silver.

Hybrid type vehicles which combine a gasoline engine with an electric motor will likely become a more common happy medium to keep jobs in place – while companies attempt to stall the inevitable transition toward an EV industry.

Remember earlier, when I mentioned that it takes about two-thirds of an ounce of silver to produce an average solar panel?
Examine how much room is left in the cost of production to demand a higher spot price in silver and still supply the world with its function.
The costs of silver relative to the price of goods from lights, hospital linens, microchips, catheters, water purifiers, vehicles, and well beyond, will have to take on the additional price increase.
We need all of our industries producing with an acceptable impact on consumers.

Inflation becomes good reasoning.
An important note to make is that electric vehicles use more silver than combustion engines.
So...an increase in the price per ounce of silver will help justify the new sticker price when electric vehicles become our new mainstream market.

And it's coming!!!

A Green Revolution will arise and eventually silver will be readily available for <u>our</u> new modern transition.
In the meantime, we will get very fortunate and watch silver explode.
But what's your definition of explode?
Mine is $84.00 an ounce versus a more intense World War. Buy before the Future Hype finds silver.
It's in our human nature to buy into Future Hype.

We believe there is value in a hefty price. Just look at Bitcoin!!!
If it happens, Fantastic! – Assuming you buy silver well before $84.00 an ounce.

Stock up in bullion and hold. You have the power when you buy and hold. That is, if you obtain it now. Everything in Future exchanges is short term for a reason!
They fund exploration ; SpaceX.
They dig banks out of ditches.
There's a pattern – Tesla – up more than 700 percent over a one-year period! Future predictions – Numerous new Tesla charging stations will arrive soon. Does that mean you buy?
Just because the train is moving fast – does not mean you need to jump on it!

Take a moment to reflect. Seven hundred percent – times it by 3 and then use common sense! Perceived value bets.

If 50 to 60 billion ounces of silver has been mined on Planet Earth throughout history and there is a generous estimate that only 6 billion ounces of aboveground silver exists –
And that estimate matches gold's existing 6 billion ounces aboveground.
That means there is a worldwide silver shortage folks!!
Let's prove it with basic math –
Let's say geographically speaking there are only a fraction of areas suitable to host solar powered homes across the world. Let's set that fraction at <u>a very low estimate of 600 million homes</u> (on a planet currently hosting 7.9 billion people) – <u>able to utilize solar panels as its energy provider.</u>

Now…take the average worldwide annual production of silver at an estimated 803 million ounces annually and times that figure by 7.5 years of Full Production Mining and Recycling and you will get roughly 6 billion ounces of aboveground silver, - matching a very generous current aboveground estimate of six billion ounces of silver in existence now.

Keep in mind that the accounting is utilizing <u>all current aboveground estimates of silver PLUS 7.5 years of Full Mining Capacity across the world mixed with recycling</u>, to reach a grand total of 12 billion aboveground ounces.

Now...We all know that silver reserves are spread across the world, so we know that the estimated current six billion aboveground ounces are spread out.
We also know that 7.5 years of worldwide mining and recycling <u>in its entirety</u> is not going to go directly into U.S. hands.

So...it's safe to say that this hypothetical scenario is not possible – unless of course my theory of U.S. hoarding silver is reality.
But let's just use 12 billion accessible aboveground ounces as an example to see what is possible with that figure and to prove we have a worldwide silver shortage <u>for a FACT</u>!

Now let's use the very low estimate of 600 million potential future solar-powered homes worldwide – (keeping in mind – all shapes and sizes of those homes) and then...conservatively placing an average of 30 panels needed per home.

Using the example of two-thirds of an ounce of silver needed to produce <u>one</u> solar panel, and that would put a rough estimate of 20 ounces of silver needed to supply that 30 panel average per home.

That means 600 million homes, at an average of 20 silver ounces needed per home, equates to 600 million times 20 silver ounces and you get exactly <u>12 billion ounces needed!!!</u> That estimate does not include businesses and all the in-betweens.
It's an estimate based on using geographically favorable locations and very low estimates to show you that funding a <u>Global</u> Green Revolution cannot happen until we create a panel that requires less silver!
Perhaps we are already there and only the U.S. knows it! Perhaps California is a test trial. Perhaps the years of hoarding silver will fund <u>our</u> Green Revolution while the rest of the world scrambles for scraps-
Or...in a perfect world, a USA led <u>Global</u> Green Revolution will arise.

Bottom line is, – with the current estimates, and the current amounts of silver needed to produce one solar panel, – we have a silver shortage globally!

Remember when I recited the information regarding <u>all the silver discovered thus far </u>would fit in a cube 55 meters on a side?
Please sit with that 55 meters example and imagine <u>two</u> ninety foot cube monuments looking at you, (basically three times higher **each** than Stonehenge's sarsens block), and you can begin to see the Silver Masterpiece unfold before your eyes.

Examine a Silver Proof Eagle coin and Lady Liberty's gesture toward the Sun. <u>Our</u> Green Revolution Prophecy.
Divine Guidance.
Does that mean you will get filthy rich by investing in bullion silver?
Define filthy rich.
Doubling or tripling your money isn't bad*
A Green Revolution isn't bad*
Getting there can be pretty ugly and take some time.
Herd mentality keeps us all enslaved.
Line on up- it is Peace.

Obtaining materials to keep the slaves working on a global movement is what makes us all a globalized world.

Assembly, materials/components, technology, design and distribution.

It's a chain gang. Industrial slavery.
The supply chain is Vital for all infrastructures and all businesses and it's vital for human lives.
If you mess with the price of silver too much – you screw with the chain. The world suffers!
My friend Josh words it poetically!
"Sometimes it's enough to know what not to do."
Do we really want to repeat history!?
Is there a choice!?

The Fed will hold the monopoly reserves as they always have.
They will control industries and markets as they always have.
They will control countries and currencies.
They will control humans.
And in essence, create as close to Peace as we can get with 7.9 billion people and counting minus Pandemic casualties.

The other central banks around the world will fall in line after the fog settles and If they're smart, well before the bombs drop.
The details and foresight of our forefathers makes me feel privileged and grateful they prepared decades in advance for war.

It puts the induction of The Coinage Act of 1965 into perspective.

Franklin D. Roosevelt once stated that the most powerful <u>remaining</u> nations will be the ones with the best soil.
Remember when I mentioned 55% of the total worldwide silver is found in just 4 countries on Earth?
Eventually, soil and water will be on the final list of wars.

If you dig deep enough, you will find the USA-owned subsidiary interests own those industries as well – all across the world!

The problem with perceived arrogance is that even after planning a counter attack 60 years ahead of our times-
When we talk about remaining soil and water, currency wars will look petty in comparison!

I believe the Defense Department sees those future wars coming to our Ports, Farms and Water Sources, and unfortunately, after those wars are over, only the elite will be able to relocate to Mars.

In the meantime, if our opponents want to spark a currency war, we will let them do the honors by pressing the restart button.

When they press that button – or when we must press that button, it will result in a more serious Global Depression, hopefully followed by a USA led Green Revolution –
And in a perfect world, humans will evolve*

The 2008 crash was strategic.
The "Bailout" Plan was drawn up prior to the crash.
Just like the Patriot Act was drawn up prior to 9/11.
Dig deep on dates before you discount these facts as "Conspiracies".
The truth hurts.
We were approaching another crash in September of 2019. A Pandemic revealed itself shortly after.
To keep Power there are many costs involved.

We are getting an inside look at how many lives are at stake with our collective compliance.
These inside looks are glimpses of what weaponized behavioral psychology can create.
A change in our understanding of definitions.
Our perception and definition of "Freedom" has changed and it is reflective in our fearful nature and culture.

Adaptive behavior changes our views on choices.
We choose what is presented to us in our environment.
Our phones choose for us. What's your frequency?

Have you ever chosen to turn your phone off for a few days and not check it? Who controls whom? We are creatures of habit.

Forty hours a week times roughly 50 weeks a year until you are 67 if you're one of the "lucky" people. One out of four people have less than $30k in their savings account and depend on Social Security benefits to survive. You played the rigged game your whole life and that's your reward. One in four men over the age of 64 is working or seeking work. One in four odds if you're lucky!

Collective Compliance.

Working Class America. Survival incentives. Our new model.

Beats a Chinese Model*

"Middle class" America in 2008 became Working Class America.

What is your definition of middle class income? $35,000 to $100,000 per year?

What is your definition of working class income? $20,000 to $60,000 per year?

Our definitions matter when you spend your whole life working. Especially when you trust your IRA! Let's examine some patterns leading up to the 2008 panic.

In September of 2007, The Dow was 13,113.38 due to worries over a global growth slowdown.

In September of 2008, The Dow was 10,365.45-
In March of 2009, The Dow hit a low of 6,469.95.
There's a new pattern in play now –

Fast forward to September 2019 and our U.S. debt accumulation went from slow and steady to rapid, and created a Phase Transition Pattern. We need to examine The Dow and the NASDAQ and what the patterns look like – and match it to news of worries over a global growth slowdown and then factor in a worldwide Pandemic to reveal what's
behind the curtain this time around.

Back in March of 2018, the Dow was at roughly 24,000. (It took 9 years to go up 17,531 points from March 2009's low of 6,469)!

Trump announced tariffs on Chinese solar panels and hit them with $50 billion in total on Chinese imports. Then Trump announced tariffs
on an additional $50 billion on Chinese imports with a minuscule trade war gesture-slap.
This resulted in China cheapening its currency in hopes it could offset some of the higher costs imposed by those tariffs and ultimately created tensions in relations.
The Dow was up to roughly 26,800 by September of 2019. A Pandemic revealed itself shortly after.

That's when the Phase Transition Pattern kicked in big time!
Time to print some money and keep afloat.
In November of 2019 The NASDAQ posted a record high of 8,386.4–

As of September 2021, The NASDAQ was roughly over 15,000.
As of September 2021, the Dow was roughly at 35,000!!! The Dow closed at 20,188 on March 16th, 2020! <u>So a year and a half to rise roughly 15,000 points!</u> <u>On 11/8/21, The Dow closed at 36,432.22.</u> <u>The NASDAQ closed at 15,982.36.</u>
Recently "recovered" markets from worries over the Omicran variant puts the Dow at 35,719.43 and the NASDAQ at 15,686.92 at closing on 12/7/2021.

This Drop and Recovery Pattern is reflective of our Short-Term Memory and Long Term Consequences Pattern, otherwise known as a Denial Pattern*, a pattern that seems to be sticking for now. We shall see how it unfolds.

To get the broader picture of how every sector will take huge hits in this phase transition pattern we are witnessing – we need to revisit mortgage-backed securities and watch portfolios rise to understand the timing of the next market crash. Greed shows its hand openly.

Pattern positioning can be seen in future forecasts. Let's use Fannie Mae and Freddie Mac as just one sector to watch portfolio wise-

Add another $100 billion plus each portfolio and we will be getting closer to Dejavu.

The Federal government was interested in Mortgage Securities and in privatizing Fannie Mae and Freddie Mac prior to the 2008 crash.

Bear Stearns paid the price – along with many others.

History repeats itself.

Patterns repeat.

As of September 30th, 2020, Fannie & Freddie had retained equity capital of approximately $21 billion and $14 billion respectively.

Note: November 2020, Fannie Mae's mortgage portfolio was $163 billion, and Freddie Mac's was $193 billion.

Update: January 2021, the U.S. Treasury establishes no exit from conservatorship with less than 3% capital.

This is called Pattern Positioning.

In January of 2018, my wife and I received a letter from Fannie Mae.
It was a "Notification of Assignment, Sale or Transfer of Your Mortgage Loan."

Fannie Mae informed us that our balance had been transferred by Wells Fargo Bank to Fannie Mae.
They went on to say the ownership of our Mortgage loan to Fannie Mae has not been publicly recorded;
Think about that transfer and their role and ask yourselves – why?
Answer: Because their role is to turn them into Mortgage-Backed
Securities.
Guess how that will unfold!?
Guess what's going to happen to JPMorgan and their silver?
Dejavu Folks!
On a Grander Scale.

Japan's Nikkei 225, Hong Kong's Hang Seng, Our Dow, Bitcoin, Gold, Ukraine, Fannie and Freddie shuffle plays, Europe and Russia's natural gas battle, Iran, ETFs, the South China Sea, Afghanistan, new Variants,
trade conflicts, the Taiwan Strait, Mining, Agriculture, Technology and the NASDAQ, all will play a role in a Worldwide Market Crash. The list goes on and on.
Don't forget the Stealth Bomber-Silver-silently awaiting release.
Let's not forget the most important unseens – Solar Flares?! Or perhaps…Our Mother Earth-
Who can create a 20% drop instantly in markets with a crushing wave or quake? We have our continued warnings.
They are luxuries.
We will look back and miss these "good ol' days!"

So here is a friendly reminder to be extremely grateful for an environment that allows you to access and obtain tangible silver without slaving for it.
Just imagine what one roll of American Silver Eagles could provide for a worker in Nepal.
That's the equivalent of two years hard labor wages!
Will we see a classic standard silver-to-gold ratio of 16 to 1 someday?

The answer is – it's very likely.
If we're fortunate, it may hold.
Gold will have to be a lot lower for that scenario to happen. Maybe $1,350.00 an ounce on gold and $84.00 an ounce on silver.

It's possible a reverse pattern could form as the herd towards gold approaches $2,500 an ounce.

This type of pattern is a momentum trap that lures into a hold and then a plummet-creating captive gold investors scratching their heads – as a reversal in gold and a rise in silver occurs. This example could create major new moves into the silver market* The silver spike pattern could surface with a plummet in gold – or a low in gold.

We are not in a conventional setting anymore.
Silver spike at silver low – could become silver spike at gold low.

Patterns break or new patterns form when there are monumental disruptions. And disruptions here we come!
The more sour silver news there is – the more incentive to buy! Silver up on gold low – it's possible*
Audrey Hepburn worded it perfectly:

"Nothing is impossible...the word itself says – I'm possible."

In 2010, forty-four million Americans were on food assistance.
That was eleven long years ago.
The same year the Mint increased its Silver Eagle production!
That year, the Agriculture Department reported an estimated 133 billion pounds of food was lost/thrown away at the retail and consumer levels.
That was almost one-third of our nation's food supply!
The Census Bureau reported about 56.7 million people in the U.S. had a disability.
How we define disabilities is going to be reflective of our overall perception of "mental health."

How does our collective health look in 2021?
We are over-due for a shift away from fossil fuels.
We know this.
We have the capabilities to transition away now.
We do not need to rely on nuclear power, coal or fossil fuels.

We can utilize what has already been mined and join forces with our Sun, Wind, and Water – to create a Green Revolution.

A quiet admission of acceptance of no control is needed before taking on a mission as grand as conquering old, stale industries. Geologists have estimated we have drilled far less than one percent into our planet's 6,378-kilometer radius – yet we've managed to alter the lifecycle of our planet and have created the depopulation of rural areas and forced mass migrations.

By 2050, it's estimated that 6.3 billion people worldwide will live in   cities. Poor air quality is a major issue now. If the estimates are correct, it would mean that over the next 28 years, the equivalent of 200,000 people worldwide per day will be moving to cities.
It also means that if air quality issues are to be remedied, we will need to form new patterns to create sustainability.

Renewable energies and smart mobility will create economic opportunities and new jobs*   District Cooling, Biomass Thermal Plants, Smart Thermostats, Automated Lighting, Wastewater Heating Spaces, Electricity-Generating Water Distribution Networks, Subsoil Energy, SMART Buildings – and Beyond will become the new norm;

By 2024, the smart home market in the USA is expected to generate over $150 billion dollars!

Some conservation technologies that create cleaner energy include Tidal Energy Generators, Fuel Cells, Hydropower, Wind Turbines, Geothermal and Silicon Photovoltaics.

Somewhere in this equation – silver patiently awaits transmutation.

The World Bank's long-term prediction forecasting a drop in silver by 2030 may be spot on. Like I previously stated, this is a marathon investment. Patience is a virtue. The Comex shows a pattern over 45 long years of a spot price average of $20.72 per ounce.

Patterns Change.
Patterns Break.
Change is Inevitable.

Only God and our attunement to our truths can help us in creating
enough bumps through our actions to enlighten us enough to create positive change.

Use your power to Boycott.
Use your power to invest in something tangible that can create optimum change*

How can we be guided if we can't listen without a device!?
Does your technology connection exceed your nature-time connection?
Silver reflects light*
Tune in to the light that silver can bring*

Get outside the box and tune in to a Green Revolution.

Back in 2013, my brother gave me two American Silver Eagle coins.
On that fateful day, unbeknownst to him, he inspired me and sparked my future interests and studies of silver.

I want to express my gratitude to my younger brother for unknowingly inspiring me to purchase silver.
I'm sure his nephew will thank us both much later in life, perhaps long after we are gone, for his planting the silver seed*

We leave our footprints with or without participation – and ignorance is bliss.

When I was a young boy, I often read Shel Silverstein's "The Giving Tree."'
I encourage you to read it again and again.

I look at Silver as a Giving Tree, and from an investment standpoint, I feel what has already been cut/dug has been done – within "reason."
I apply it to how I invest in silver and pay the extra premium for it.
It helps me sleep better at night – considering what mining does to our Earth and to our people.
Label me a hypocrite.
You live with silver – look around.
Money makes the world go round.
Lousy justifications!
We are all slaves in some sense and we support slavery with our choices. Period.
Does it make it less evil to influence or inspire potential investors in silver to pursue silver mined 20 plus years ago?
No!
It's easily a karmic copout!
But I'd like to encourage potential investors not to participate in the current models of production.

I invest <u>strictly</u> in what I like to call Numismatic Bullion*

Meaning older year bullion like 1986 American Silver Eagles. Currently priced between $37.00 and $45.00 per coin. If you do your homework! I buy them for my safe batch.

That means they are safely stored for my children's children.

I'd suggest a <u>sell</u> <u>batch</u>, (when silver reaches $84.00 an ounce), of older 100 ounce silver bars. A simple online search can educate you and lead you towards older bullion – preferably older Eagles, and older produced bars.

Creating a work such as SilveRevolution, comes with great responsibility for me personally.

It goes against my common sense ideals of Clean Water, Clean Air, and Preservation of Land.
It goes against my interpretation of morally acceptable.
It can easily be misinterpreted.
In layman's terms, I want to see people prosper and apply that prosperity towards sustainable living.

We can evolve in our decision making.

We can influence markets with our purchasing or lack of purchasing.
We have the capacity to choose right from wrong.
Humane Certified vs. Industrial Caged.
Solar Power vs. Nuclear Power.
Love vs. Fear.
Locally Owned vs. Corporate Owned.

Our Roots are Fearless in Nature.
With our choices, I believe we can motivate industries to adapt to less silver consumption, and eventually, we can inspire new solar panel development. We can buy and hold.
We can create a Movement –
Call it SilveRevolution!

We still have our purchasing power and it speaks volumes!
We can Boycott on a larger scale.

I'm not an advocate for the Silver Industry.
I appreciate Silver's value in its role to help the world in regards to optimum energy production and less pollution.

I'd buy nothing produced after 2001 for a reason – Because we have an estimated six billion ounces of aboveground silver and we still have a Giving Tree left!
I do not want to encourage us sitting on a stump at the end of the book – like the character in Shel Silverstein's book.

I'd like to encourage you, if you have been inspired to purchase silver, to only purchase silver produced decades ago and in moderation within your accessible savings account*

No sense in purchasing new pain and suffering.
We have plenty of that with a Pandemic.
Here's to a Green <u>Revolution</u>*
Even if you lose, we can collectively profit*
I don't know of many other investments that can provide that!

With the perception of "freedom of speech" and the "right" to an opinion, - I should underline that my sources for material and facts regarding all context in this book could be considered as my perceived beliefs based on what I've read about silver over the years.

As far as my selected sources and bibliographic articles of references go, I will state for the record that my Silver Prophecy is my opinion.

I invite everyone to research what I've presented and form your own opinions.

In definition, that's the beauty of freedom of expression...or...what's left of it.

A mandate by definition is the will of voters as expressed in an election.

The birthright of our bodies, and our laws of free will to do what we please with our bodies, does not apply.

Your compliance will change the definition of Freedom and we will all ultimately live with what is left as our choices. Or…we will attune to our light and love and create a nurturing environment.

Acknowledgement is a vital key in optimum growth. Somewhere in the tree rings of others, your acknowledgement is present.

Even if you received no response from your acknowledgement, it is silently creating. All too often we discount the creative expressions presented to us daily. Modern times make it very easy to overlook our art in being.
Sharing our moments and receiving acknowledgement deserves more than a like icon or a pat on the back or verbal great job!

It deserves genuine interest, compensation, and encouragement to inspire additional sharing. Today, I acknowledge you for your services.* You are a creative being helping others create their rings in growth.

Growing upward like the great redwoods, you touch the light and move your branches so the light can touch others down below.

You are appreciated*
I appreciate that you have taken the time to read this book and support the art of expression.

ACKNOWLEDGEMENTS

I give thanks to my loyal friend and music companion, Gary Slaughter, for deciphering the relays and creating tangible works, including the artwork for this book.*

My friend, Saint Andrew deserves an honorable mention for his consistency, patience, and thoughtfulness in admiral fashion and for helping publish this book.

My friend, Brian Sendler, deserves recognition for contributing photography for this book.*

****

In closing, I'd like to thank my Mom for lovingly typing and for selflessly supporting me my entire life.*

I'd like to thank Judith and Davlo for their great patience, love, support and sacrifices in helping my family.

I'd like to thank Sharon and Gary for providing the best education a child could hope for!

Thank you to Mother Earth for enduring her hosts. Thank you to all the friends, family, and Silver enthusiasts/experts, for the Great Disillusionment Gifts.
They are Priceless!

After 20 years of marriage, I thank my wife for teaching me to hope for the best and "prepare" for the worst. Thank you to my son, Jacob, who sets an example we should all follow by being genuine love always. A true inspiration and motivating force!

As my friend Sean always says, "Get rich quick and count your many blessings."

God Bless us All*
God Help us All*

silver-evolution.com